NORTON COMMANDO

ALL MODELS

Roy Bacon

NITON PUBLISHING

First published in the United Kingdom by:
Niton Publishing
PO Box 3 . Ventnor . Isle of Wight PO38 2AS

Acknowledgements
The author would like to thank those who helped this book by supplying photographs. Most came from the EMAP archives, or *Motor Cycle News* by courtesy of Editor Malcolm Gough, and some from the Author's own archives. My thanks to all.

© Copyright Roy Bacon 1989

Filmset and printed by Crossprint
Newport . Isle of Wight

ISBN 0 9514204 3 7
A CIP catalogue record for this book is available from the British Library

All rights reserved. No part of this publication may be reproduced, stored in a retrieval system or transmitted in any form or by any means, electronic, electrical, chemical, mechanical, optical, photocopy, recording or otherwise without prior written permission from the publishers to whom all enquiries must be addressed.

Front Cover: The front cover is taken from a 1970 Norton brochure and shows the S model with its lovely exhaust pipes.

Back Cover: 1974 850 Roadster

A very early Commando, then listed as the model 20M3, during a trip to the International Six Days Trial in 1968.

Contents

Introduction	4
The Launch Commando	6
Into Production	24
The 850 Commando	42
Road Racing and Specials	48
The Interpol	52
Specifications	54

Introduction

The Norton Commando was produced between 1968 and 1977 following a September 1967 launch. It was built in a dozen different forms over the years but all used the same twin-cylinder engine derived from that designed by Bert Hopwood back in 1948. That had first appeared as the 497 cc Dominator but had been stretched, in stages over the years, to 745 cc for the Atlas and it was this form that went into the Commando and, in time, was enlarged still further.

The feature that set the Commando apart from the rest was not the engine but the mounting system that held it, along with gearbox and rear pivoted fork plus wheel, to the frame. The design was given the name Isolastic and it insulated the rider from the engine vibration, although it did nothing to repress this. Much else of the design was common to all versions of the Commando with the variations coming mainly from the cosmetics of tank, seat, panels and bars; but items such as gearbox and forks had their roots in the past.

The Commando frame with its Isolastic mounting system was unique and designed as a means of preventing the bad vibrations of the engine from reaching the rider. This problem had gradually become worse as the capacity and engine speed increased and reached its peak with the 745 cc Atlas first seen in 1962. It is an inherent feature of any 360 degree vertical twin and the solution of isolating the source from the rider was certainly ingenious, even if it did nothing to actually remove the cause.

The system was conceived by Bernard Hooper and Bob Trigg who quickly developed the idea into hardware. The isolating rubbers were worked on to bring the vibration level down and, in the end, little could be felt above tickover. When combined with the well developed engine, the result was one of the fastest machines then available for the road.

The Commando became a truly classic motorcycle which remained as popular in the 1980s as when first launched. Despite some problems, large numbers continue to enthral their owners by the sheer grunt of the engine allied to precise steering.

The 1968 model 20M3 with the inclined engine and separate gearbox of all Commandos. The massive footrest plates never break.

The Launch Commando

The Commando made its debut at Earls Court and attracted immediate attention with its new frame, silver finish and strange green roundels on the tank sides. These, it was explained, were the new group motif and signified worldwide use, but most visitors concentrated on the machine itself.

The construction concept was that the engine, gearbox and rear fork were assembled as a single unit with front and rear engine plates. This unit was then fitted to the frame with principal mountings to front and rear which incorporated isolating rubbers. A further rubber mounting acted as a head steady and the silencers, being rigidly connected to the engine by the exhaust pipes, were also attached to the frame via rubber mountings.

Everything else was either fixed to the frame or the engine and where to both, was flexible, as with the pipes and cables. The front and rear mountings did not permit totally free movement of the engine unit but controlled it, so that it was mainly along the line of the machine with limited side play. What there was of this was set by shims at first and riders soon learned that the setting affected the tautness of the han-

Drive side of the Commando when first seen and thus with a normal multi-plate clutch under that well polished chaincase.

THE LAUNCH COMMANDO

The Commando as first seen with its new engine mounting concept designed to keep engine vibration away from the rider.

dling and the degree of vibration that did get through the system. They also learned that the settings had to be regularly checked and adjusted if necessary, as too much clearance gave poor handling and none at all would break the frame!

The original design had included a vernier adjustment to make it easy to alter the setting but this was not adopted until 1975, for cost reasons. The shims were cheaper, as was the detail design, and riders were to curse the awkward but essential service task.

The Commando engine was very much derived from the 1948 Dominator and a number of details were common to both. It was separate from the gearbox and, before the Commando, was always installed with the cylinders upright. For the new frame and model it was inclined forward a little so that the cylinders were nearly parallel with the downtubes, giving a new lease of life to the engine line.

The engine was effectively that used in the Atlas and its 745 cc capacity came from a 73 mm bore and rather long 89 mm stroke. The construction was in the British style with the aluminium crankcase castings split on the engine's vertical centre line. Each carried the same size main bearing but that on the drive side was a roller, while that on the timing side was a ball race. An oil seal went outboard of the roller race and an oil sealing disc beside the timing side bearing.

The mains carried a built-up crankshaft which comprised a central flywheel with a crank half on each side with the parts clamped together by four bolts and two studs with a single dowel to assist alignment. Each crank half was a single

7

NORTON COMMANDO

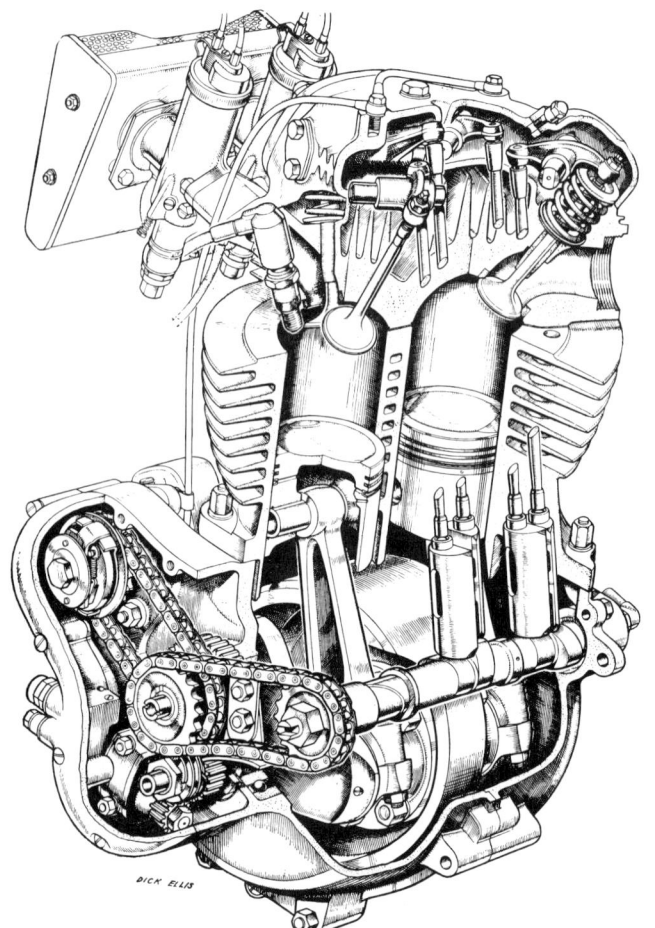

The Atlas engine which went into the Commando with minimal change other than to the inlet tracts and ignition system.

forging of mainshaft, web, crankpin and inner flange. The crankpins were hollow to provide a sludge trap and drilled for lubrication with cross holes to the big end surfaces.

The crankcase assembly had the cast iron block held to it by a ring of studs and a selection of nuts. There were two pushrod tunnels in the front of the casting and the hollow tappets were located at the base of these, where they ran in the cast iron. A small plate went between each pair to prevent them rotating in their common guide hole and kept them in place when the block was lifted.

Within the block the three-ring pistons moved in step and were handed by the valve cutaways in their crowns. They were attached to the connecting rods by hollow gudgeon pins, themselves retained by circlips. The rods were forged

THE LAUNCH COMMANDO

in light alloy and the gudgeon pins ran directly in the small ends, but there were replaceable shells for big ends. Each rod cap was held in place by two bolts, each with location diameters and an eccentric head to locate and lock it in place. Stiff-nuts completed the assembly and had a listed torque setting.

There was a gasket under the block and another between it and the head, the top one being in solid copper for the early Commando engine. The cylinder head was an aluminium casting with integral rocker box and was fixed to the block by an array of studs, nuts and bolts. Some ran up into the head and others down to the block, which minimised their effect on the fins, the air flow over them and thus the cooling. It did, however, make removal of the head an awkward job with the engine in the frame and not the easiest even when it was on the bench.

Still the model 20M3 with the extended dualseat that was one of the styling features from the start.

NORTON COMMANDO

Bernard Hooper points to the front Isolastic mounting on a Commando during the presentation of the Castrol award the design won.

The head itself was much as it had been in 1948 with widely splayed exhaust ports and twin, parallel inlets, with some downdraught. This was offset by a separate, curved, inlet tract for each cylinder which brought the twin 30 mm Concentric carburettors back close to the vertical, the two tracts being joined by a flexible balance tube.

Each valve guide was pressed in to a flange and its valve was controlled by duplex coil springs retained by a collar and split collets. There were cups under the springs and heat resistant washers between these and the head itself. The rockers were positioned and formed so that they conveyed the movement of the four pushrods, which ran up tunnels cast in the front of the head, to the valves.

The rockers carried adjusters

THE LAUNCH COMMANDO

Commando on show in 1968 with the seat extended to form kneegrips and the fibreglass tail section of the early model.

with locknuts at their outer ends and each was forged in one with a hollow spindle. Each oscillated on a fixed pin which was a tight fit in the head and sealed to it with a pair of bolted on plates, the inner one of which prevented the pin from rotating. Access to the valves and rocker adjusters was via caps, with one each for the exhausts and a single cover at the rear for both inlets.

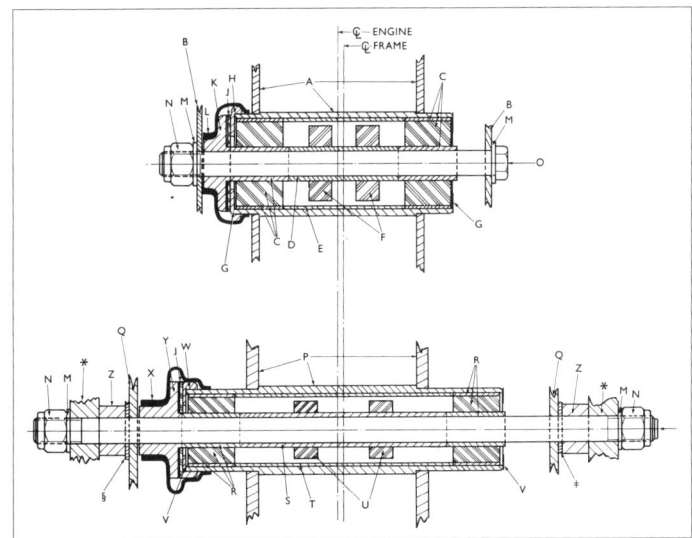

The Isolastic mountings which set the Commando apart from the rest with the front one at the top, the rear below.

NORTON COMMANDO

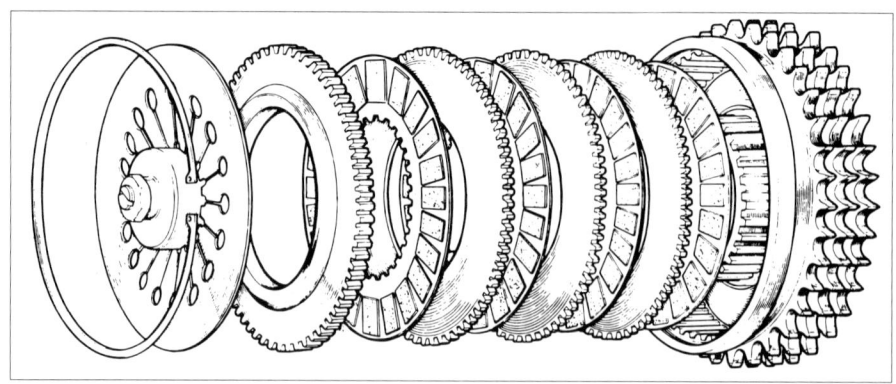

Details of the diaphragm clutch adopted for production and made for Norton by Laycock Engineering. Do not dismantle without the right tool.

Rocker lubrication was by an external pipe from the pressure side of the oil system. The supply was taken from the timing chest to two banjo connections in the top of the head and then by internal drillings to the working surfaces. The oil finally drained down the pushrod tunnels and back to the crankcase. This system improved the earlier, rather marginal, method of supply from the return line, but did require the pipes and joints to be up to coping with the high supply line pressure.

The exhaust pipes were held to

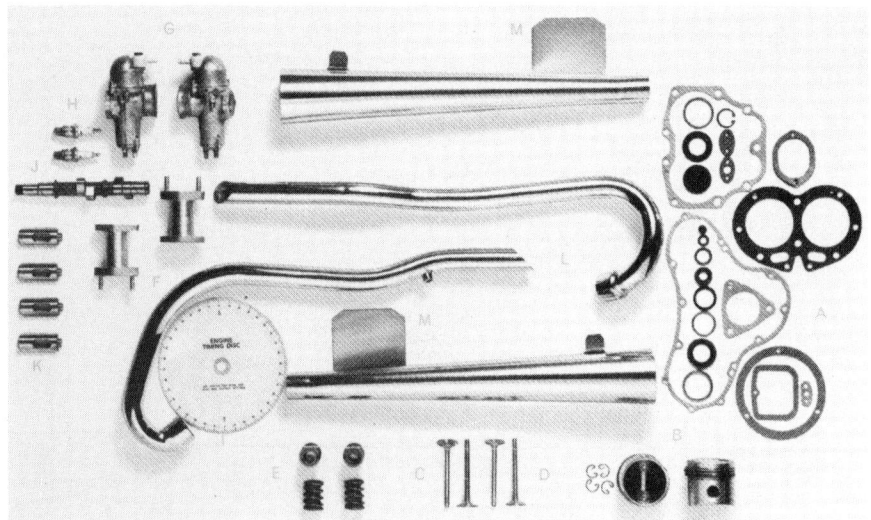

Stage 3 of the tuning kit which took the machine to 137mph, courtesy of Paul Dunstall, who developed all the kits.

THE LAUNCH COMMANDO

Fastback MkII of 1970 with the timing chest points and revised rev-counter drive but still with gaitered forks.

the cylinder head by large finned nuts which were screwed into the ports. The nuts were always something of a problem to keep tight, which was important as, if loose, they could knock the port threads out. These can be repaired but only at some expense.

The camshaft ran across the front of the engine in bushes in the crankcase halves and was driven in two stages by gear and chain. The first stage comprised a pinion on the crankshaft meshed to an interme-

Commando Production Racer on show in 1970 with tuned engine and cockpit fairing.

NORTON COMMANDO

During 1969 the first Commando was renamed the Fastback and this name came to be applied to all the machines with this style of seat.

diate gear which ran at half engine speed. This part incorporated two chain sprockets and ran on a fixed spindle pressed into the crankcase. The sprocket chains ran fore and aft, the front, outer one driving the camshaft sprocket, which was keyed in place. This chain was adjusted for tension with a small slipper working on its underside.

The second chain drove back to a twin contact breaker unit, with an advance mechanism, which replaced the magneto of old and was

A Commando S on test in 1970 and sweeping past a Messerschmitt bubble-car.

THE LAUNCH COMMANDO

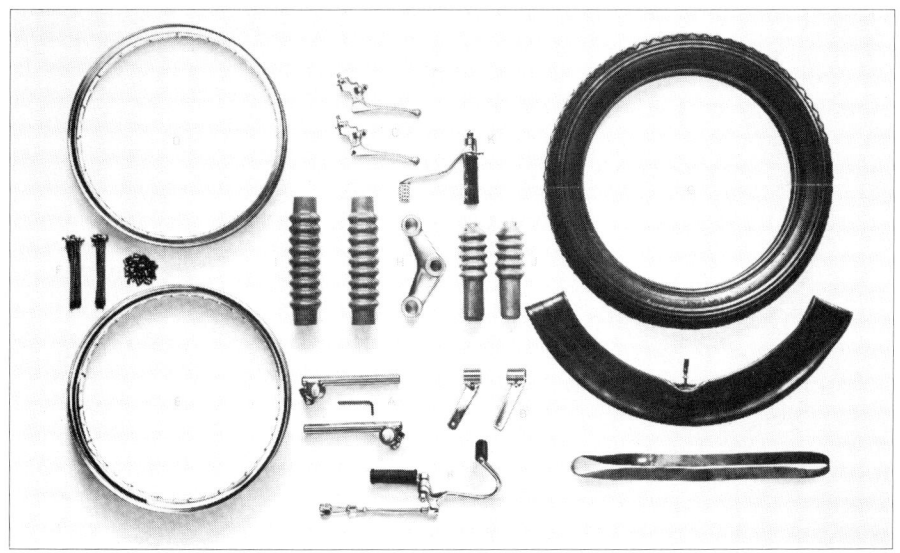

The custom kit also offered in 1968 and also never to reach the public.

tucked in behind the crankcase and block. This chain was adjusted by moving the complete points housing. Ignition was by external coils used in conjunction with a capacitor that was powered by the Lucas RM21 alternator fitted to the left-hand end of the crankshaft, outboard of the primary chain. Its output was controlled by a zener diode in conjunction with a rectifier so the ignition system was independent of the 12 volt battery.

The left-hand end of the camshaft drove a timed disc-valve breather with an elbow pipe outlet screwed into the crankcase. The breather passage ran down the centre of the camshaft to a point between each pair of cams where cross-drillings communicated with the engine interior.

The right-hand end of the cam-shaft drove a rev-counter gearbox fixed to the outside of the timing cover. This last item enclosed all the timing gear and the oil pump, which was driven by a worm that was screwed on to the crankshaft to hold the timing pinion in place. The pump fed its output into the timing cover via a sealed bush. The cover carried the pressure release valve and from this the oil flowed to a chamber in which was located an oil seal running on the crank-shaft end. This ensured that the oil went into the shaft and then on to the two big-ends.

The oil pump was a duplex-gear type as had been used by the firm since 1931, but the assembly was modified for the twin. The pump had both pressure and scavenge gear pairs to suit the dry-sump system and it worked well, except for a

NORTON COMMANDO

tendency to let the oil drain through it into the sump when left to stand.

The Commando used the standard AMC gearbox, with a revised shell and internals, but kept the stock layout and change mechanism. It was driven by a triplex primary chain, which made alignment much more important than with the earlier single strand, and for the Show launch it used the conventional six-plate clutch, clamped up by three springs, and incorporating a shock absorber.

The gearbox design dated back to 1935 when Norton took it over from Sturmey-Archer, improved a few detail points and had it made for them, by Burman. Years later, in 1956, after Norton had become part of the AMC group, the box had the positive stop mechanism revised along with the clutch and the result was used for all heavyweight Norton, AJS and Matchless models. It was known as the AMC type and continued with minor changes on the Commando.

The primary drive was enclosed by a large alloy chaincase which carried the alternator stator on bosses cast as part of the inner. A single nut held the two halves together and was sealed by a large O-ring set in a groove in the inner case joint face. The arrangement worked well as long as all the castings had their distance pieces in place, for without them it was all too easy to crack something.

The engine, gearbox and chaincase assembled as one unit, with front and rear mountings. The first of these simply comprised two small plates with a large tube welded between them and this assembly was

The S model Commando with its lovely exhaust on the left of the machine, this being a 1970 model.

THE LAUNCH COMMANDO

The 1970 Roadster which was essentially the S with low level exhaust pipes and silencers.

bolted to the front of the crankcase. The rear was similar, but larger, as the plates again bolted to the crankcase, but then extended back round the gearbox. The mounting tube was welded in place at the top rear corner of the plate and beneath that went a second cross-tube for the rear fork pivot pin.

The pin extended on each side

The Fastback MkIII of 1971 with slimline forks but otherwise as the II and still with the same rather odd dualseat.

NORTON COMMANDO

Custom style Commando Hi-Rider of 1971 with its strange seat, handrail and excessively raised bars.

of the mounting plates so that the fork, with its flanged bushes, could pivot on it. The assembly included felt seals and end caps so that the hollow pin could act as a reservoir for oil to lubricate the bearings. Unfortunately, it was not to be very successful, as owners tended to use grease instead of oil, which blocked the minute lubrication holes, so the bushes ran dry and then wore rapidly.

With the rear fork attached to the mounting plates the complete

Outside the works with a Fastback III, Fastback Long Range, another MkIII and an Interpol from left to right.

THE LAUNCH COMMANDO

assembly could be offered up to the frame. This looked conventional, and in most ways it was, with a single top tube and twin downtubes which ran back under the engine and gearbox; the tubes then turned up to join the subframe loop and support the tops of the rear units. A bracing tube ran forward from each, in the vicinity of the rear fork pivot, to join the top tube halfway along its length. Initially, the headstock was braced to the top tube with a gusset, but this proved to be a weak point and within a year there was a bracing tube beneath the top one.

The engine assembly was attached to the frame front and rear with similar rubber isolating assemblies. Essentially, each was a spindle bolted to frame lugs, which ran across on the centre-line of the tube welded between the engine plates. Between the spindle and tube were two sets of rubber discs, one, in soft rubber, being a fit to both, and the other, a harder compound, which fitted the spindle but had clearance to the tube to act as a bump stop. Distance tubes and shims controlled the clearance from side to side, and there were bellows to enclose the ends, seal them and keep the lubricating grease in place.

Above the engine went a head steady which was bolted to the rocker box and allowed some movement, while keeping the whole assembly upright in the frame. As the assembly included the rear wheel,

For 1972 the engine was revised and is here installed in a Fastback MkIV.

NORTON COMMANDO

it was of some importance, but the original type proved prone to fracturing. It was eventually redesigned, but not until 1973; most early models have since been fitted with the later part. When correctly set, the Isolastic system worked well and, in time, riders, especially production racers, learned to tune it to suit their needs. It was also found that, while the shims could be adjusted to reduce the working gap, it was essential to have some gap as, without it, the parts became solid, a condition which could break the frame and possibly the head steady as well.

At the front, the frame carried its own version of the short Roadholder telescopic forks with hydraulic damping, internal springs, gaiters to protect the main tubes and short top covers with headlamp lugs. The front wheel had a pull-out spindle and a full-width, light-alloy hub with 8 in. twin-leading-shoe brake, which had been an option for the earlier Featherbed twins. Its backplate had an air-scoop and three round air outlets; a rod link connected the

The engine unit as for 1972, still inclined forward and fed by twin Amal Concentric carburettors.

THE LAUNCH COMMANDO

Fastback LR or Long Range, of 1972 and thus a MkIV, with increased tank capacity and revised seat to accomodate the result.

two cam levers and there was a separate cam and pivot for each shoe. The hub was spoked into a steel rim and the wheel shod with a 3.00 x 19 in. tyre.

The rear wheel matched the front with regard to the hub but had a 3.50 in. section tyre and 7 in. single-leading-shoe brake. This worked in a drum which incorporated the rear wheel sprocket and was held to the hub by three long sleeve nuts

Interstate on show, with the new silencers of short and long megaphone sections, and a Production Racer behind it.

NORTON COMMANDO

The 1972 Production Racer, still with cockpit fairing and the yellow finish which made it easy to pick out at a distance.

screwed to integral studs. Thus it retained the quickly-detachable feature of old.

Both wheels were covered by sports mudguards but that at the rear was mainly out of sight under a tail unit. This glassfibre moulding ran back from the seat to continue the styling line and supported the rear lamp and number plate. Beneath the seat, on the left, there was a side panel but on the right this space was occupied by the oil tank. Between tank and panel sat the battery and other electrical incidentals which were all controlled by an ignition switch mounted just ahead of the side panel on a frame bracket.

Above the panels went the dualseat which was orange on the show model but became black in production. It was held in place by a large hand-nut on each side and was notable in having ears that ran forward on each side of the petrol tank to act as kneegrips. The tank itself was in glassfibre with a snap-action filler cap and held just over three imperial gallons of petrol. The combination of tank, seat and tail worked well.

Below the side panel and oil tank there was a forged aluminium support plate on each side. Each carried a footrest hanger which ran forward to the footpeg itself with the left one also supporting the rear brake pedal. The plates extended back to carry the pillion footrests

THE LAUNCH COMMANDO

and also provided a pick-up for the flexible silencer mountings.

The exhaust pipes ran low down on each side to tubular silencers that lacked a tailpipe and were common to other twins in the range. A single air filter was connected to the two Amals and had a replaceable element. The assembly fitted in between the oil tank and side cover but was not tailored to match them at all.

A matching speedometer and rev-counter were provided, each in a housing held by a fork top nut. They too carried the anonymous green blob which also went on the tank sides and tail unit, to confuse rather than clarify. The controls were conventional, with an air lever combined with the front brake lever and both centre and prop stands were provided.

For the Show, the finish was mainly silver, the frame, tank, tail unit, side cover and oil tank being in this colour, along with the cylinder block. The mudguards and headlamp shell were chrome-plated and the primary chaincase well polished.

The machine was well received.

Into Production

The new machine was first listed as the 20M3 model, this becoming its engine and frame number prefix. At the launch there were brochures listing custom and conversion kits, the former for added style and the latter for extra performance.

The custom kit was more in the style of a café racer for it included clip-on bars, rearsets, alloy rims and alloy racing clutch and brake levers. The top yoke became alloy, and there were coloured gaiters for the forks and the rear units, new headlamp brackets, an Avon Grand Prix rear tyre and a transfer for the rider's helmet.

The conversion kits provided three levels of tune and resulted from collaboration with Paul Dunstall, who was by far the most successful Norton twin tuner at that time. The first stage raised the compression ratio to 10:1 and included a pair of long tapered megaphone-shaped silencers to push the speed

Roadster MkIV from 1972 with its short, sharp looks, pleated seat and conical ends to the silencers.

INTO PRODUCTION

up to 120 mph. Stage two brought in a new camshaft, exhaust pipes, inlet tracts and inlet valves to go to 130 mph; while stage three added a hotter camshaft, racing exhaust system with megaphone, bigger carburettors and a further 7 mph.

The clutch really was new, for it had a diaphragm spring to clamp the plates, then common in cars but not motorcycles, other than those with the Villiers Starmaker engine. All parts of the clutch were new and, as the design precluded

Final 1973 745 cc Roadster in MkV form complete with turn signals, disc front brake and old type silencer.

All the kits included various other odds and ends, but in the end none went into production.

The Commando did though, in April 1968, and by then it had a new clutch and a degree scale on the alternator rotor to assist ignition timing checks with a stroboscope.

the fitting of a shock absorber in the centre, the transmission had to manage without this aid to comfort. The clutch was designed by Laycock Engineering and had a high clamping pressure with reduced lever effort.

The production finish was more

NORTON COMMANDO

sober than at the Show with the frame and cylinder block in black, oil tank and side cover remaining in silver, a black seat and the petrol tank and tail unit in green. This proved just right and the machine was an immediate success, enhanced by it winning the *Motor Cycle News* 'Machine of the Year' poll in 1968. It was to go on to repeat this one to prevent breakages. The model then took the name Fastback, due to its tail unit which was to remain a feature to its end. It was joined by two further models, the R and S, which dispensed with the tail so had a stock rear mudguard.

The R was aimed at the North American street scrambler market and had a smaller petrol tank in red

The MkV Interstate of 1973 when the 745 cc engine was over its troubles and one of the best to buy.

for another four years, to make five in a row, up to 1972.

The machine was still known as the Commando model 20 Mark III and it stayed that way until March 1969. By then the block colour had reverted to silver and the frame had been modified to include the extra bracing tube under the main top or blue, conventional dualseat and high bars but otherwise was as the Fastback. The S had more style for the same market with twin exhaust pipes that curled round to waist level on the left. They were coupled to reverse-cone silencers mounted one above the other and both pipes and silencers had perforated heat

INTO PRODUCTION

The Interstate which first appeared in 1972 with large petrol tank, disc front brake and an engine that was about to run into trouble.

shields to protect the rider and passenger.

The style did not stop there for the S had slimline forks without gaiters, which did nothing for the life of the seals or stanchions, and a chrome plated protector ring round the headlamp shell. The seat and tank colours were as for the R but there were two side panels. Both had a smaller, more triangular, shape and the left-hand one carried

A Commando built up by Challenge Motors of Barcelona in 1972 with a nice combination of stock and special fittings.

NORTON COMMANDO

Peter Williams having the works Commando refuelled during one of his Thruxton rides.

INTO PRODUCTION

Norton owed much to his fine engineering and racing abilities.

NORTON COMMANDO

The Hi-Rider was built with the 829 cc engine and retained its seat and bars but did have the disc brake fitted.

the ignition switch.

The S was the first model to have a modified engine design with the points moved into the timing cover where their cam was driven from the end of the camshaft. This deleted the awkwardly placed distributor and its chain and meant that the rev-counter drive had to be moved. For this, a skew gear was cut on the camshaft and the drive taken from this, the housing being fitted to the top of the crankcase.

Some of the R models also had this later engine fitted, but this remained a rare machine which was soon dropped.

Up to July 1969 the Commando was built in the old AMC factory at Plumstead, marque production having been moved there from Bracebridge Street, Birmingham, in 1963. Shortly afterwards, the building was compulsorily purchased for later demolition so the firm had to move again and the logical site

INTO PRODUCTION

Mkl 829 cc Roadster of 1973 which kept the style it had always shown but with the disc front brake.

would have been the Villiers factory in Wolverhampton. However, government inducements persuaded them to set up the assembly line at Andover, Hampshire, although manufacturing did go to the Villiers plant where the engines and gearboxes were put together. They were then transported the 120 miles to Andover overnight and this split build arrangement was to produce many problems over the years.

One of the better aspects of the Andover site was the proximity of Thruxton race circuit where the test department was set up. Later came a performance shop to make and sell special performance parts under the Norvil label.

In September 1969 the Fastback went over to the later engine with the timing cover points and, together with the S, it went forward for 1970 in this form. They were joined by the Roadster in March that year; this was essentially an S with low-level exhaust pipes, and all models were offered in a variety of colours for petrol tank and side covers. In the middle of the year the S was dropped and in September 1970 the Fastback became the Mk II when it was fitted with the pipes and reverse cone silencers from the Roadster.

This was only an interim move for January 1971 brought the Fastback Mk III while the Roadster became the Mk II. Both had many detail changes including the slimline forks without gaiters. There were also alloy control levers with the electrical switches incorporated in their pivot blocks but these Lucas items were still far removed from the ergonomically correct designs

NORTON COMMANDO

A 1973 Roadster fitted with the 749 cc short-stroke engine plus performance parts to make it a fast road or competition model.l

that had become common on Japanese machines. There was a new oil tank and a right-hand side cover for the Fastback to conceal it. The head bearings were changed to sealed ball races which worked very well as they could not be adjusted incorrectly and the grease stayed where it was needed. A shock absorber was built into the rear wheel between the hub and brake drum, smoothing the transmission while retaining the quickly-detachable feature.

The two models were joined by four more in the next few months and the first to appear was the Street Scrambler in March. This had a waist-level exhaust on each side that terminated with the reverse-cone silencer. It had a small headlamp and petrol tank, sump guard, braced handlebars and sprung front mudguard. Its tank colour was yellow or tangerine but it was only listed for a few months.

Next came the Fastback LR and the Production Racer. The former was simply the stock Fastback fitted with a larger-capacity steel petrol tank and a seat to suit, while the initials 'LR' stood for 'Long Range'. The latter was rather different and much more serious, the prototype having been built in 1969. This single machine was followed by more in 1970, these being built by the Norvil performance shop.

INTO PRODUCTION

For these machines the compression ratio was raised, another camshaft and bigger valves fitted, and all the usual performance tweaks carried out. These worked well, for the Norton cylinder head design was excellent and lent itself to this work. On the cycle side there were clip-ons, rearsets and a cockpit fairing, all of which enabled a Commando to win the Thruxton 500 mile race that year.

There was also a long, and expensive, list of optional fittings which began with a disc front brake, included four- and five-speed close-ratio gear clusters and continued with all the expected race-type details. All these added up to a formidable motorcycle that proved well able to hold its own in production and 750 cc racing. The 1971 Production racer was finished in yellow, so it was inevitable that it became known as the 'Yellow Submarine' or 'Peril', the first relating to the Beatles' song of that name. It was supplied with all the quick bits plus a Lockheed disc front brake to stop it.

Nice 1974 MkIA Roadster with the bean can silencers and black air filter box but no fork gaiters.

NORTON COMMANDO

John Player Norton version of the Commando with its stylish seat and fairing, matt black exhausts and fine lines.

The final new 1971 model was called the Hi-Rider and was a chopper-style custom model built to pick up sales from that area. The bulk of the machine was pure Commando with the small petrol tank and side covers. To suit the style, the model was fitted with high-rise, but unbraced, handlebars and a strange dualseat with high hump that made

The Formula 750 Racer as listed in the 1973 brochure.

INTO PRODUCTION

two-up riding nearly impossible. A grab rail ran up just aft of the hump to complete a machine that was far removed from the normal Norton image.

For all that, the Hi-Rider continued into 1972, when there were some extensive engine changes, and the Fastback, LR, Roadster and Hi-Rider all became Mk IV models. They were joined by a new model, the Interstate, which was built as a tourer and had a petrol tank size in excess of five gallons and new silencers. These had a short, shallow taper megaphone and long reverse-cone, also with a shallow taper. In addition to the road models the production racer continued in the range.

The 1972 engine had new crankcase halves with internal webs to increase the main bearing support. The timing side main became a roller race to match the drive side and, thus, the crankshaft had to float between rollers, with shims to limit the end play. The timed breather was replaced by a separator that was bolted to the back of the crankcase, low down, and had a pipe connecting it to the oil tank. A cartridge oil filter was added to the oil system and was fitted between the engine and gearbox in the return line.

The engine was also available built to a higher degree of tune and, in this form, was called the Combat; this being an option for the

Another view of the JPN showing the twin headlamps set in the fairing which incorporated a facia for the instruments.

NORTON COMMANDO

The Thruxton Club Racer, with 749 cc short-stroke engine, which was briefly listed for 1975.

Roadster and Interstate. The extra power came from a raised compression ratio achieved by skimming the cylinder head, a hot camshaft and larger carburettors. The block was painted black on this engine. In addition to the engine option there was also one for a disc front brake which was listed for the same two models. It was based on that used for racing but with the caliper on the right, behind the fork leg.

It said 'Electric Start' on the side panels even if it was something of a misnomer and this is the installation of 1975.

INTO PRODUCTION

A MkIII Interstate on show with left side gear pedal and disc rear brake among it many changes.

In either form, the engine lost its sump filter so the scavenge pump was no longer protected, had a weak advance mechanism to drive the points cam and continued to run rather too fast thanks to the gearing employed. Both wheels had changed to 4.10 x 19 in. tyres and the standard sprockets resulted in only 107 mph at 7000 rpm engine speed, which all could pull easily.

Naturally, owners used the performance, so most Commandos spent a lot of time at a high engine speed. This gave the tired old crankshaft design a hard time for it had been laid out a quarter of a century before to cope with 500 cc, 29 bhp and 6000 rpm. The increases to 750, 65 and 7000 were all too much and it began to flex, or as much as its new stiff crankcase would allow.

The conflict became concentrated on the main bearings, where the inner races followed the crank, so all the load was applied to one section of the track and the roller corners, which tried to dig in. The problem was compounded by the timing chain often not being in adjustment, because this was awkward to do, and once slack it was subject to odd loads and reversals from the hot camshaft. This played havoc with the ignition timing which was further upset by the ineffective advance mechanism, that would often vary wildly or jam on full ad-

NORTON COMMANDO

vance.

The upshot of all this was a very short main bearing life and a great number of warranty claims. The answer was the Superblend bearing which was a roller race with each roller having a slight barrel form at each end. These could cope with the flexure and speed, while in time, the ignition advance mechanism was improved. This was not all, however, as the high engine speeds brought out a weakness in the pistons which were slotted below the ring belt. These were fine up to 5000 rpm, but above that speed the crown could separate from the skirt, after which the debris would pass through the oil pump to do a complete wrecking job on the engine.

Norton realised that they could not continue in this way so decided to revamp all their stock and drop the Combat engine, despite having announced that, together with the disc brake, it was to become the standard, except for the Hi-Rider. The decision cost them dear, for every machine had to go down the line to be dismantled and again for re-assembly, while the engines went back and forth between Andover and Wolverhampton.

Norton 76 being shown off with its cast alloy wheels, twin front discs and SU carburettor, but only the one was built.

INTO PRODUCTION

Close up of the works Norton engine installed in the multi-tube frame tried by Peter Williams at one time.

NORTON COMMANDO

Commando customised by Gus Kuhn in 1971 to match the mould of the time.

All engines had the new mains and much improved advance mechanism while the machines had the gearing raised by two teeth on the gearbox sprocket. That proved much better, as only racers could pull that ratio to 7000 rpm in top and they had none of the problems thanks to their frequent careful rebuilds. This left the high compression ratio on the Combat which proved awkward to reduce as there was no way of replacing the 1 mm that had been machined off the head. An extra thick aluminium gasket was tried but found wanting as it compressed, while two thinner copper ones were difficult to make gas-tight and usually leaked oil.

All told it was a bad year, but matters improved a good deal for 1973 for the final 745 cc engines proved to be good ones. The range did not appear until March when the two Fastback models were dropped but the Roadster, Interstate and Hi-Rider ran on in Mk V guise. These retained the 32 mm carburettors of the Combat engine and all had minor detail changes and the option of turn signals, which had only been offered for the Interstate the previous year. The production racer continued to be listed in its yellow war-paint and was joined by a Formula 750 Racer, for

INTO PRODUCTION

events to those regulations. For these, a racing fairing and seat unit were fitted but still in the yellow.

During 1973 the firm became dragged into the whirlpool of events at BSA and Triumph from which Norton Villiers Triumph, or NVT, emerged. Later in the year came the announcement of plans that resulted in the famous Meriden sit-in and Norton finances became another of the political pawns on the board.

While this went on the works built their final 745 cc Commando models in October but the replacements were already in the market place and had been there since April 1973.

The 850 Commando

The final stretch of the original Dominator twin engine came early in 1973 when the 745 cc engine was bored out to 77 mm and 829 cc. Three models were offered with the same names as the smaller ones, but there were enough changes to make the two types easy to tell apart.

For the engines there were new cylinder blocks with fixing bolts that screwed down into the crankcase. To enclose the bolts the block casting was extended out to produce a distinctive form beneath the fin area. Inside the engine went a flywheel of suitable weight and the sump filter made a welcome return. Engine breathing was much improved by using the primary chaincase as a vent chamber and taking an outlet from the top rear corner of this.

On the outside, a balance pipe appeared between the exhaust pipes close up to the ports; while there was a host of detail changes to the cycle parts. The disc front brake went on to all models and only the Hi-Rider was left with a glassfibre petrol tank. The silencers continued to vary with the reverse cone type on the Roadster and Hi-Rider and the short megaphone, long reverse cone type on the Interstate. There was a range of colours listed for the various models and twin lines went on the tank and side covers to distinguish the larger engines.

There was also a new silencer type which appeared in the brochure on a Roadster and an Interstate, both being fitted with an enlarged air filter box in black. These were the Mk 1A models which came to the lists in September 1973 for the European market and offered a reduced noise level. The silencers had a megaphone taper leading into a long body with matt black end mutes and quickly became known as the 'bean can' or 'black cap' silencers. They worked well, reduced the noise without affecting the power, so soon became the recommended wear for all.

The brochure included pictures and information on a special short-stroke engine which was offered as a unit by itself or installed in a Roadster. The reduced stroke brought the capacity down to 749 cc, while retaining the 77 mm bore, and the specification included most of the performance parts. Thus there were high-compression pistons, big valves, hot camshaft, steel rods and electronic ignition. The Roadster with this engine was built to the usual form to offer a high performance road model or, it was suggested, the machine could be a base

THE 850 COMMANDO

for further work for competition. To this end the firm offered megaphone exhaust systems and 33 mm carburettors with bellmouth intakes to push the power up to a claimed 80 bhp at 8000 rpm.

To round off the brochure there were photographs of the works production and Formula 750 racers in the livery of their sponsor, John Player tobacco. Both were built as low as possible in an endeavour to reduce frontal area, to compensate for a lack of power.

The association with the sponsor led to the appearance of the John Player Norton model late in 1973 although it did not reach the shops until the following April. The machine was based on the standard model to which were added styling features much on the lines of the production racer. The engine itself was the stock 829 cc unit, the 749 cc short-stroke being an option. It was fitted with a high-output alternator and an exhaust system finished entirely in matt black including the 'bean can' silencers.

It was the bodywork that set the JPN apart for most of it was special and it included a fairing complete with screen, instrument panel and twin halogen headlights; hence the need for the high-output generator. The fairing was in white with styl-

This nice 829 cc MkIII Interstate was pictured in 1983 being raffled by the excellent Norton Owner's Club which is recommended to all.

NORTON COMMANDO

ing stripes on each side and swept back to hide the cylinders but not the crankcase or primary chaincase.

The fascia was moulded-in and carried the speedometer, rev-counter, three warning lights and a switch. Mirrors and turn signals were fixed to each side and behind the fairing went a white, moulded tank shroud to continue the line. The shroud was taken out at its base to bolt to the fairing to brace it and provide a top to the cooling-air tunnel. The tank itself went beneath the shroud which provided the mounting for the snap-action filler-cap that was hinged from it.

The seat was a single one with the padding fixed to a tail moulding that extended back to carry the rear light. This tail was also in white, as was the moulded front mudguard but the rear guard remained in steel. To go with the café racer style there were clip-on bars and rearsets which picked up on the usual holes in the hanger plates. The brake pedal continued to be part of the left footrest assembly but the gear pedal was simply reversed on its spline.

The Mk IA Roadster and Interstate models ran on for 1974 while the Mk I versions became the Mk IIA with the new air-box and 'bean can' silencers plus a pleated seat top. Some of the Mk IIA machines delivered to the North American market retained the earlier air-box and silencers for the air-box definitely reduced the performance. The Hi-Rider changed to the Mk II but retained the older air filter and

First year for the 829 cc engine was 1973 and this is the Interstate fitted with it and the usual large tank for its MkI form.

THE 850 COMMANDO

A glance back at the start in September 1967 when the Commando was first shown to the public, complete with its strange green tank markings.

reverse-cone silencers.

The NVT saga was by then in full flow and of considerable complexity due to the many factions involved. Changes in government, the energy crisis, the three-day week and inflation all played their parts in a situation that was complex anyway. BSA had gone and Meriden was still occupied by the workers' sit-in, so Norton had to do the best they could.

For 1975 the range was reduced to two road models and there were several major changes which included electric start, a left-hand gear pedal, disc rear brake and hinged seat. There were many more detail changes and the most obvious was the move of the front disc brake to the left, with the caliper ahead of the fork leg. This improved the handling for abstruse reasons clear to no-one. On the service side the appearance of the vernier adjustment for the engine mountings was the best news, as this made the task of adjusting the gaps much easier.

NORTON COMMANDO

The Mk III mountings could be adapted for earlier models and some owners took advantage of this.

The pedals were switched over to suit the American buyers and the brake one, plus a hydraulic master cylinder for the rear disc, went on the hanger plate easily enough. The gearchange was less simple and the solution adopted was to retain the entire change mechanism, but with the pedal shaft exiting inboard from the end cover instead of outboard. A cross-shaft was then coupled to it and run through the chaincase to the pedal but via a pair of spur gears.

This made the change heavier and required the gearbox to take a fixed position which left primary chain tension to deal with. For this, a twin plunger device, which worked on both chain runs, was adopted and went into the chaincase. This was further modified to take the starter motor and its drive train, to the left-hand end of the crankshaft, but this device never amounted to more than an assistance, so it was as well that the kickstart remained. Finally, the chaincase halves were held together by a row of small screws, rather than the old single nut, and the seal became a gasket.

It added up to an improved motorcycle which was offered in Road-

The Dunstall export model, one of the many varieties that Paul could ring up by changing seat or tank and adding a fairing and other useful fitments.

THE 850 COMMANDO

The 1974 Interstate MkIIA with black cap silencers, new air box and pleated seat top.

ster or Interstate Mk III forms. For Europe the forks even had their gaiters back again but seat, side cover and petrol tank were the only real differences between the two models. With them was the Thruxton Club racer fitted with the 749 cc short-stroke engine and to be built in very small numbers indeed for Formula 750 racing.

The two road models ran on into 1976 while the commercial and financial areas underwent further twists. It all became even more complex but by the end of the year a batch of 1500 machines had been laid down, after halts and crises for many months. The batch was not finished until late 1977 after which a few more were assembled at Andover in 1978. Prior to then, the works had built a restyled Commando with cast alloy wheels, twin front disc brakes and a single SU carburettor. This was named the Norton 76, from the year, but only the prototype was made.

The final Commando year was really 1977 although one more machine surfaced as late as 1982. Since then the factory has been stabilised, after many difficulties, to build machines with a rotary engine. This has revived the name in the showroom, while the Commando became linked to the classic machine growth of the 1980s.

The Commando had many irritating faults but at heart offered the essence of the British motorcycle with pulling power and the ability to ride far and fast in a comfortable manner. Many are still out there doing just that and in regular use, supported by a very active owners' club.

Road Racing and Specials

These two are linked, as often one led to the other. The Norton twin began to be raced in the early 1950s but it was not until the next decade that it achieved much success. During this period the factory ran a racing twin at Daytona and the TT with some good outings.

The first machines were called Domiracers and ran well but the larger 646 cc versions did better, especially in production racing. The knowledge gained went into the 745 cc Atlas in due course so, by the time the Commando came along, the necessary tuning data was well established.

Much of it had been first learnt by Paul Dunstall who had tuned and raced a Norton twin in the late 1950s before turning to race preparation at the age of 21. He quickly became the prime supplier for all manner of special parts for the twin and late in 1962 he was able to buy up the Norton race shop when it closed. This brought him the works Domi-

Works Norton with one of the many chassis variants used under the guidance of Peter Williams who combined his great riding skill with less weight and height to offset a lack of power.

ROAD RACING & SPECIALS

Dave Croxford riding the Commando in typical style - hard and very, very fast.

racer and a host of spares which helped him as he sponsored riders through the 1960s.

In time, Dunstall's road machines became accepted as a recognised make and, in 1968, won the production TT in the Isle of Man. He used the Norton Featherbed frame for most of this time but, as the engine size and power rose, found he needed something else so used Rickman, Lyster and his own design of spine frame in the latter part of the decade.

By then, he had become involved with the Commando tuning kits and quickly had his own line of parts for the model. These were both cosmetic and performance orientated and included an 810 cc kit

NORTON COMMANDO

which numbered an alloy block among its parts. As before, Dunstall went on to offer complete machines in various guises from tourer to production racer, and often these incorporated innovative features.

Gus Kuhn Motors of London followed in the Dunstall mode with special parts and a production racer, for they were a major Norton dealer. In time, they also offered complete

ing and being responsible for much of the development. This resulted in the production racer, first listed for 1971, but available 12 months earlier, built up from an extended range of options. Norton won the Thruxton 500 mile race in 1970, Williams being one of the co-riders, and he came very close to winning the production TT, fuel starvation putting him under two seconds behind the winner. He went on to

The 750 Production Racer for 1973 as shown in the brochure for that year.

modified machines in a choice of specifications, mainly in the café racer mould. Both Kuhn and Dunstall continued this activity into the early 1970s, but then moved on to other things.

The factory started to take an interest in racing the Commando in 1969 with Peter Williams both rid-

several good performances in 1971, which clearly showed the potential of both machine and rider.

This resulted in the formation of a factory team for 1972 with John Player tobacco sponsorship, but this had limited success. Daytona and the TT brought more troubles than places, although Phil Read man-

ROAD RACING & SPECIALS

Stylish travel for the team with smart and well equipped Dodge transporter.

aged fourth in the USA and Williams a second in the Island. Both positions testified more to the skills of the riders than the professionalism of the team.

They did win the Thruxton 500 once more and for 1973 travelled to their races in an excellent transporter based on a Dodge chassis. Peter Williams finally won his much merited TT, the Formula 750, with Mick Grant bringing the other Norton home second, but there was little else to bring the factory joy.

By then, the firm had become interested in using a twin-cylinder Cosworth engine for their racer, this being essentially a section from the very successful grand prix car V8. For this, Norton planned a monocoque frame for 1974, but all these attempts to overcome a lack of power came to nought with the advent of the big three- and four-cylinder racing two-strokes from Japan.

Gulf Oil were the sponsor for 1974 but the firm had little success and during the year Williams crashed badly, never to race again. The sponsors withdrew and 1975 saw the end of any works racing. The Commando continued for some time in club production racing and, in America, one machine was very successful in the flat-track races held at Ascot Park in the mid-1970s. The Cosworth engine did get built and the machine ran in one race but that was all. The Commando's race day had run.

The Interpol

Neale Shilton built up enormous goodwill and a great rapport with police forces in many countries while working for Triumph. When he finally left that firm he was, in time, recruited by Dennis Poore to create a police machine for Norton. He joined that company in 1969.

Internal company problems prevented the first from being completed until 1970 but the result proved satisfactory and the machine, named Interpol, began to be ordered by police forces at home and abroad. Not too much of it differed from standard and most that did was peripheral. Thus there were white mudguards and side panels,

One of the series of Interpol models developed by Neale Shilton, this one with an 829 cc engine.

THE INTERPOL

a single seat and a choice of petrol tanks, one with a recess for a radio. Both were in steel, to suit legislation, and based on the Atlas tank with a new base.

There was a certified speedometer which went behind a fairing whose screen carried the police blue lamp. There were panniers for equipment and two-tone horns for pulling road users over for that friendly chat, but the essence was the basic Commando.

Shilton made great efforts to make the Interpol succeed but was often frustrated in this by others in the firm. The Commando was basically an excellent machine for police work with enough power and a relaxed manner, which made it easy to use for a day-long shift. These good points had to be allied to correct assembly and reliability of both machine and spares supply, but these aspects were all too often missing.

The Interpol was built right through to 1976 so some came in the Mk III form with its many alterations. This meant that it went through the trauma of the Combat engine, despite Shilton's efforts to prevent this, and he writes that one batch of 25 sold to Kuwait finished up in the Arabian Gulf, dumped due to overheating.

Not quite the end anyone wanted.

NORTON COMMANDO

Commando Specifications

All models have two cylinders, overhead valves, Concentric carburettors and four-speed gearbox.

Model	750	750	750	850	850	750
years	1968-71	1972	1973	1973-74	1975-77	1973
bore mm	73	73	73	77	77	77
stroke mm	89	89	89	89	89	80.4
capacity cc	745	745	745	829	829	749
comp. ratio	9.0	9.0[1]	8.5	8.5	8.5	10.5
carb type	930	930[2]	932	932	932	933
top gear	4.85	4.85	4.38	4.38[5]	4.18	4.38
front tyre	3.00x19	4.10x19	4.10x19	4.10x19	4.10x19	4.10x19
rear tyre	3.50x19	4.10x19	4.10x19	4.10x19	4.10x19	4.10x19
front brake dia	8	8[3]	10.7[4]	10.7	10.7	10.7
rear brake dia	7	7	7	7	10.7	7
wheelbase in.	56.7	56.7	57	57	57	57

[1] - 1972 Combat engine 10.0
[2] - 1972 Combat engine 932
[3] - 10.7 disc option for Roadster and Interstate
[4] - not Hi-Rider
[5] - 1974 4.18

SPECIFICATIONS

Commando Specifications

Models built were :

745 cc				829 cc			
Model	Years	Tank gal		Model	Years	Tank gal	
20M3	4.68-3.69	3.25		Roadster I	4.73-12.73	2.5	
R	3.69-9.69	2.25		Roadster IA	9.73-2.75	2.5	
S	3.69-6.70	2.25		Roadster IIA	1.74-2.75	2.5	
Fastback	3.69-8.70	3.25		Roadster III	2.75-1977	2.5	
Fastback II	9.70-12.70	3.25		Interstate I	4.73-12.73	5.5	
Fastback III	1.71-12.71	3.25		Interstate IA	9.73-2.75	5.25	
Fastback IV	1.72-3.73	3.25		Interstate IIA	1.74-2.75	5.25	
Fastback LR	4.71-12.71			Interstate III	2.75-1977	5.25	
Fastback LR IV	1.72-3.73			Hi-Rider I	4.73-12.73	2.0	
Roadster	3.70-12.70	2.25		Hi-Rider II	1.74-2.75	2.0	
Roadster II	1.71-12.71	2.25		JPN (850/750)	11.73-2.75	3.5	
Roadster IV	1.72-2.73	2.5		Thruxton racer	1975	3.5	
Roadster V	3.73-10.73	2.5		Norton 76	1976		
Street Scrambler	3.71-10.71						
Production racer	4.71-10.73	3.5 or 5					
Hi-Rider	5.71-12.71	2.0					
Hi-Rider IV	1.72-2.73	2.0					
Hi-Rider V	3.73-10.73	2.0					
Interstate	1.72-2.73	5.25					
Interstate V	3.73-10.73	5.5					